Richard Kerr

Wireless telegraphy

popularly explained

Richard Kerr

Wireless telegraphy
popularly explained

ISBN/EAN: 9783741191350

Manufactured in Europe, USA, Canada, Australia, Japa

Cover: Foto ©Andreas Hilbeck / pixelio.de

Manufactured and distributed by brebook publishing software
(www.brebook.com)

Richard Kerr

Wireless telegraphy

PORTRAIT OF MR. W. H. PREECE, C.B., F.R.S.

Wireless Telegraphy

POPULARLY EXPLAINED

BY

RICHARD KERR, F.G.S.

AUTHOR OF "HIDDEN BEAUTIES OF NATURE"
AND PUBLIC LECTURER ON EXPERIMENTAL PHYSICS, ETC.

WITH A PREFACE

BY

W. H. PREECE, C.B., F.R.S.

NEW YORK

CHARLES SCRIBNER'S SONS

743-745 BROADWAY

1898

Preface

———

THIS little work, although extending only to about one hundred pages, contains a good account of the discoveries in telegraphy without connecting wires. The subject-matter is arranged in a readable form, the illustrations are excellent, and the descriptions of the experiments are accurate.

Mr. Kerr having visited Dundee, the town where James Bowman Lindsay lived, and where his memory is greatly and very justly respected, has evidently been caught in the whirl of local enthusiasm. Lindsay was undoubtedly a man of great originality of ideas, but some of his theories on electrical signalling were not novel in 1854, and were deemed impracticable at the time. I was

the officer appointed by the Electric Telegraph Company to assist him in making
his experiments in London.   As a scholarly
linguist, however, he, by unaided effort,
accomplished great things in literature, quite
sufficient to make him famous.

In taking account of the early work done
in this department of science, the earlier experiments of Morse which were carried out
by Gale in 1842 on the Susquehanna River, as
recorded in Vail's early work on telegraphy,
should not be forgotten.   Several other
American professors and electricians have
recently had considerable success in their
investigations of conduction and induction
as applied to signalling without intervening
wires — notably Professor Trowbridge of
Harvard, Mr. Wiley Smith of Kansas, Mr.
Tesla, and Messrs. Phelps, Gilliland, and
Edison.   In England we succeeded in
bridging the Solent in 1882, and in 1896
in communicating with the Fastnet Lighthouse by Mr. Willoughby Smith's conduction method.   In India communication
has been maintained across rivers for prac-

tical purposes on a conduction system devised by Mr. Melhuish.

As regards Signor Marconi's position as a discoverer, it should be understood that he was the first to conceive and to patent the application of Hertzian waves to telegraphy apart from mere signalling. Branly made the first coherer, but Marconi was unquestionably the first to make the coherer into a telegraphic relay.

My own system is now in daily practical use by the War Department between Lavernock and Flat Holm in the Bristol Channel. In my own researches I have received invaluable assistance from Mr. Gavey, Mr. Kempe, and Mr. Arthur Heaviside.

There are a great many points connected with each of the systems before the public that require to be threshed out in a practical manner before either of them can be adopted for general use. The chief use to which a system of ethereal telegraphy will be applied is most likely to be for shipping, lightship, and lighthouse purposes.

Without accepting any responsibility for the controversial points raised in this work, I am glad that the subject of electric signalling without intervening wires has been treated in so popular a form, and made easy of comprehension even by those who have but little acquaintance with electrical phenomena.

W. H. PREECE.

GENERAL POST OFFICE,
    June 30, 1898.

Contents

List of Illustrations

Introduction

EXCEPTION has been taken to the term 'wireless telegraphy' as applied to methods of signalling across space without intervening wires between the apparatus used in sending messages and that which receives them. The objection raised is that such a term is not scientifically accurate, inasmuch as wires are used in the construction of the apparatus, or in base-lines, or lines placed in parallel position for induction purposes.

But it has been adopted largely because it conveniently draws attention to the discoveries in this department of experimental physics rather than for its scientific accuracy.

At Mr. Preece's lecture in 1894 on 'Electric

Signalling without Wires' fault was found
with the title of the subject, but the chair-
man, Sir Richard Webster, Q.C., M.P., said
he thought the objection to the title of
the paper was rather hypercritical, because
ordinary people always understood telegraphing by wire as meaning through wires
going from one station to the other, and
these parallel wires not connected would
rather be looked upon as the sending and
receiving instruments. He hoped, therefore,
that the same name would be adhered to in
any further development of the subject.

These chapters contain the substance of
a lecture which has been delivered by the
writer in many of the chief cities of England,
Scotland and Holland. His aim has been
to make this brief account of the details
of a wonderful subject simple and intelligible
to the unscientific reader rather than to
give it anything of the character or com-
pleteness of a text-book.

In the appendix will be found certain definitions and detailed descriptions which might not be intelligible to the general reader, but may interest those who have some elementary knowledge of science.

Wireless Telegraphy

___

CHAPTER I.

SUPPOSED ORIENTAL POWERS OF SIGNALLING
THROUGH SPACE WITHOUT WIRES.

THE following remarkable instances of the mysterious powers said to be possessed by Eastern nations of sending messages across distances of hundreds of miles without visible appliances have been brought to the notice of the writer.

Three of the instances were mentioned to him by officers of the British Army; the fourth was related to him by a gentleman occupying a high position in Holland.

It appears that during the Indian Mutiny more than one system of secret signalling was in full operation. The plan of stamping cakes of bread with secret messages, and despatching them from town to town and village to village, was soon discovered, and was clumsy and slow compared with other more ingenious methods.

There were more occult ways that were harassing to the military authorities, but the secret was not discovered.

In all probability any such secret method could only be known to a very few.

During a recent war in Afghanistan, whenever the British officers conveyed to their subordinates particulars as to their intentions to operate at a certain point fifty or a hundred miles away, the natives shortly afterwards knew all about their plans. ' Had we,' said the officer, ' sent out men on the swiftest horses at our command with a view to the intercepting of the tidings of our proposed movements, they would have been too late in every instance. We watched the hill-tops to see if any method of visible signalling from hill to hill had been resorted to, but we failed to detect anything that could be of any service to us in solving the mystery. We then resolved to bribe the natives, and accordingly subscribed a considerable sum of money, which we offered them in return for a full solution of the methods they employed. But our efforts were without the result we looked for. Money could not purchase the secret from them. They seem to look upon it as a religious possession, which they must not part with to the heretics of the Western countries.'

The natives of the Dutch East Indies are also credited with a similar power. A gentleman now living in Amsterdam informed the writer that when he held an official position in one of their East Indian possessions some years ago, the natives knew everything of importance that occurred in any of the other islands situated several miles away. ' If any catastrophe took place, whether caused by natural forces, such as an earthquake or a storm involving shipwreck and loss of life, or if a murder had been committed many miles away, the natives on the island on which I lived,' he states, ' would know all the particulars long before the tidings could be conveyed by the ordinary channels, however expeditious. Later on, as a steamer would enter the harbour, those on board would naturally imagine they were conveying to us information that would be

quite new, but of which our people had
made us cognisant immediately after the
occurrence. Although I lived for several
years in an official capacity on these islands,
and made repeated endeavours to induce
them to divulge their methods, I never met
with success.'

Still more extraordinary is the power said
to be possessed by certain native Egyptians,
if the following statement can be substan-
tiated, which is vouched for by a military
officer who has had considerable experience
in Egypt : 'The day that England, and, in
fact, all humanity, lost General Gordon at
Khartoum, several of the people in the
streets and bazaars of Cairo knew of his
death.'

The question arises, How did they get
hold of the information ?

There is no railway nor telegraph system
from one city to the other, though doubtless
both means of communication will be in
evidence in a very short time, owing to
British pluck. The distance in a direct line
from Khartoum to Cairo cannot be less than a
thousand miles. Even had there been a line
of railway, and had a train been despatched

at the rate of sixty miles an hour, it would have taken over sixteen hours for the journey.

Some who have faith in clairvoyance assure us that such feats of telegraphy or telepathy are easy of performance by persons having clairvoyant power; that distance, and even the intervention of so-called solid bodies, are no hindrance to the transmission and reception of communications.

If this could be satisfactorily established, the explanation rests at once in a mental power about which little is known, as its foundations have not been traced to a generally accepted scientific basis.

Fortunately or unfortunately, there are some of us who require a scientific explanation of all the mysteries of our surroundings. This may arise from our ignorance, from sheer curiosity, or from a sincere desire to arrive at truth.

Whether our requirements will ever be granted or not in this world is somewhat problematical.

A good deal of comfort may be got, however, from the opinion of a well-known thinker, who states that ' we are yet only just making a faint scratch on the surface of the big lump of things knowable.'

It would seem that the Oriental methods of signalling without wires must rest entirely on a highly trained mental effort.

If the mesmerist, or hypnotist, as he is now called, can influence a subject far removed from him at any time he chooses to do so, and can send him into a cataleptic state, and afterwards restore him to his normal condition, it necessarily follows that one mind must have some connection with another.

Here is a department well worth investigation, but the public mind is not yet prepared for it.

Many instances might be quoted to show that surprising results may be accomplished in mental telepathy between two persons who are in thorough sympathy with each other, and who undergo careful practice in the communication of thought from one to the other. One mind transmits, the other receives impressions.

It is quite possible that the Orientals give more attention to this kind of mental training, and are therefore able to accomplish the wonderful feats of signalling attributed to them.

Everyone, at one time or another, has been made familiar with remarkable phenomena that could be explained in no other way than by one mind acting on another mind.

To say that all are the result of coincidences would be absurd.

If we do not understand certain, at present, inexplicable phenomena, let us at least be logical and maintain an open mind, and not discredit the efforts of those who know more than ourselves; neither should we attribute all such efforts to spiritualism, whatever that may mean.

We are surrounded with mysteries, while our vision and all our other powers are limited. There are forces and forms of energy, undreamt of, awaiting investigation. Many of them doubtless would be beneficial to the human race if we could only lay hold of them.

The discovery of the X-Rays is an indication of this.

We have had X-Rays present for twenty years in all experiments where high currents of electricity have been applied to Crookes's tubes, yet we never realized the fact that such an aid to medical science was so close at hand.

The whole nature, power, and virtues of the beam of light are not yet known, and so it is with many of our surroundings.

CHAPTER II.

IS THERE ANYTHING SOLID?

THERE are three or four different systems adopted by scientific men for signalling through space without connecting wires, and it is evident that if we are to understand any of them we must modify very considerably our ideas of the compactness of solid bodies.

In fact, it becomes a question whether there is any substance in nature so absolutely compact that there are no spaces between the atoms into which a thin medium may enter.

Just as our ideas of the opacity of wood, ebonite, aluminium, etc., underwent a complete change on the discovery of the X-Rays, so it is with regard to everything that unscientific people have been accustomed to look upon as absolutely solid or compact.

It is not easy to give up old ideas, but what are we to do when we find that electricity applied to a wire underground influences other wires miles away, as in many experiments carried out by Mr. Preece?

Or, again, when Dr. Oliver Lodge or Signor Marconi places a transmitter in one room and his delicate receiver in another, or removes it to a greater distance, and the electrical energy immediately passes through walls, books, and in fact everything, our ideas of matter, of solid bodies, and of compactness, receive a rude shock, and we wonder what we are coming to.

For my part, I know of nothing more wonderful than that if I press a button here a definite number of times there is at once an exact response recorded on a ribbon of paper miles away, in the complete absence of wires between this point and that in the distance!

A few simple illustrations will help us to see that bodies which we look upon as perfectly compact must have spaces or interstices between their atoms, however close we may imagine those atoms to be together.

The enormous tree of the forest, which

has grown for a thousand years, and has withstood the storms of those centuries and defied the ravages of insect life and of decay, may be looked upon as a solid body possessed of perfect compactness to the core.

Suppose such a tree to be cut down and its branches lopped off, and its length to be 200 feet, while its girth is proportionately great, doubtless it will take a great amount of steam power or manual labour to raise the giant 1 foot from the ground. It may be many tons in weight, still, it is not truly compact, for its atoms are separated by spaces, and the atoms have room to move.

If we take a pin and gently scratch one end across the grain, so gently that we cannot hear the sound of the scratching at our end, a friend at the other end will distinctly hear the sounds produced, and will be able to tell us the exact number of times we used the pin.

To do so he need not put his ear against the tree, for he can hear distinctly several inches or a foot away from the end of it.

Now, what has taken place? The gentle scratching with the pin has set every atom and molecule in the trunk in a state of rapid vibration.   The pressure of the pin may not have been more than half an ounce, yet it is sufficient to cause such an alteration in the huge mass that waves of sound are rapidly conveyed all through it.

This is a simple experiment that any two persons may try for themselves.   One of the large blocks of Norwegian timber frequently seen lying on quay walls will do equally well. To many the result is somewhat startling at first.   It may be urged that in this illustration we are dealing with *sound* waves, and that such waves have nothing to do with Mr. Preece's or Signor Marconi's methods of telegraphy, inasmuch as their experiments depend exclusively on forms of *electrical* energy.   The illustration is given to show that wood, however healthy and bulky, is not solid to the exclusion of spaces between the atoms.   We are simply clearing the decks, so that we may understand subsequent points of importance in these wonderful discoveries.

Suppose we now select something else of a more compact character.   A razor which consists of refined steel becomes fatigued from over-stropping, and requires rest.

A hairdresser who has used razors for forty years gives it as his experience that excessive stropping spoils the razor temporarily; that it should be stropped immediately after, and not immediately before, use.

Like the tree, the atoms and molecules become disarranged, and they require time for their rearrangement to take place.

This alteration in the condition of atoms and molecules is known as the 'fatigue of metals.' Metals actually get tired and require rest.

To obviate the razor trouble, wives would do well to provide seven razors, so that the husband could have one for each day in the week. Thereby they would preserve both the temper of the razor and that of the husband at the same time.

Those who have used Crookes's tubes for any considerable time will have learnt that glass becomes fatigued and requires rest.

It is customary now to keep several tubes in readiness, and to use them in rotation, owing to this very alteration in the atomic and molecular condition of the glass.

When a tube has been used it is laid aside, and after an interval of rest it works as well

as ever.   It seems strange to speak of glass
and steel becoming tired and requiring rest.

Let us consider another example of so-
called solidity.

Tin by itself is a soft mineral, so soft that
it can be cut with a knife.

Copper is soft also, but if we mix them in
the proportion of 77 parts of copper to 23 of
tin we produce a compound called bronze.
The composition is hard, and is used as bell-
metal.   It is much harder and more compact
than either of its component parts.   Still,
the alloy is wonderfully adapted to the pro-
duction of musical notes, and vibrates under
simple touches or scratches.   Hard as it is,
its atoms have room to move.   Therefore,
we must look upon it as lacking the perfect
solidity we are seeking.

But we have not considered a special kind
of bronze that is far harder than bell-metal.
In prehistoric times bronze was so splendidly
made and hardened, that certain implements
were capable of working granite, as in the
Egyptian monuments.

A fortune could be made by the re-discovery
of this process of bronze-hardening.

All kinds of proportions of the constituent

minerals have been tried, but without success. The process is an obsolete art; but the bronze, even if as hard as our steel, is not scientifically compact to an ultimate degree, inasmuch as its atoms may be disarranged in various ways.

One more illustration, on a larger scale than any of the foregoing, may help us to realize more fully that metals are not as compact as we fancy they are.

A single line of metal rails from London to the Forth Bridge is about 400 miles in length. These rails, owing to the linear expansion alone, are said to be 400 yards longer in the summer than in the winter.

If the expansion be a half, or even a quarter of this length, it is a surprising fact. But the metals expand in other dimensions as well as in length, therefore a considerable alteration must take place during extremes of either heat or cold.

The open spaces purposely left at the ends of each length of rail, frequently half an inch in width in the winter, are almost closed up in the summer. Great allowance must be made for expansion, otherwise the metals in summer would bulge either in or out, and cause railway accidents. One of the greatest forces known to scientific men is the expansion of iron on the application of heat.

In the construction of the Forth Bridge and the Clifton Suspension Bridge allowance on a tremendous scale had to be made for expansion and contraction. Such mighty structures would soon collapse if provision of this kind had not been made. They contain no more iron in summer than in winter, but the atoms are driven further apart by the action of heat. Seeing all this, we naturally ask whether even so dense a thing as steel can be called solid.

We must retain the terms *solid* and *solidity*, because they enable us to distinguish between bodies that are gaseous, or liquid, and those that are more compact; but in looking around us, among all the substances of the earth's structure, or among those produced by the art of man, we cannot find anything that is perfectly solid.

This conclusion at once prepares the way for the next statement, which will bring us a step nearer to the full appreciation of the value of the principles underlying the great discoveries first indicated by Clerk Maxwell, and experimentally detected by Hertz : *If there be nothing absolutely solid in Nature, it follows that it is possible for a medium possessing certain qualities to permeate all things.*

We have the strongest reasons for believing that such a medium exists.

The great fact to be apprehended in connection not only with the several systems of telegraphy without wires, but with many other departments of science, may be expressed in the following words : Throughout all the so-called solid materials of the earth, through all the liquids and gases, through ourselves and our atmosphere, throughout the space between our earth and the moon, through the moon itself, throughout all the vast distance of ninety-three millions of miles between us and the sun ; in fact, throughout all things too small to be seen by the power of the best microscope ever constructed, and through all the space ever reached by the largest telescope in the world, there exists a medium known as *the ether*. In fact, all interstellar space across which light travels, whether from our sun or from any other star, is filled with this ether.

If we see light from a remote so-called

'fixed star' (for no star is really fixed, all have their motions), that light received and appreciated by our eyes must have travelled through this mysterious medium, even though the star be incalculably remote.

Now we have to see how the ether is of service in the discoveries we have under consideration.

If a stone is thrown into the middle of a pond, a series of ripples, or small waves, cover the surface of the *water*. Similar waves are produced in the *air* whenever a bell is struck ; and the *ether* has its waves also.

It was conjectured by Faraday, Helmholtz, Stokes, Clerk Maxwell, and others, that *light* from the sun and *electricity* were the same in *kind*, and that they only differed in *degree*, the difference resting in the lengths of their respective waves. Their velocity through space was the same, namely, 186,400 miles a second.

Later, Dr. Hertz actually proved in his laboratory that electro-magnetic waves were capable of reflection, refraction, and polarization ; also, that the electro-magnetic waves were longer than those of light.

This paved the way for surprising results, some of which will be noticed later.

Energy sent out from the sun receives different names. For example, we have *light* waves, *heat* waves, *electric* waves, and so on. These are all undulations of the ether.

But all these forms of energy, in travelling from the sun to the earth—a distance so great that an express train travelling sixty miles an hour without stopping would take 175 years to accomplish it—reach our earth in eight minutes.

The waves cannot travel along nothing. They must have an elastic medium which will transmit them.

Without attempting to fathom the many mysteries of the ether and its remarkable properties, we may feel sure of its existence.

If, therefore, it be capable of conveying energy from the sun — say, for instance, electrical energy—' without loss ' (to quote Mr. Preece), and without intervening wires, it is reasonable to ask, Why cannot we devise some form of instrument that will also send out along the all-permeating ether electrical energy, even in a small way ?

2—2

The ether will act as the medium, and electricity as the messenger.

We have, then, but to devise some sensitive instrument which will receive a share of the energy thus sent out.

It is a matter of surprise that Professor Joseph Henry, of Washington, when he had made a wonderful discovery nearly sixty years ago, did not follow up the result of his own experiment.

He was using a current giving a spark of 1 inch in the top room of his house, and he succeeded in setting up induced currents in wires in his cellar.

These currents, whether the result of induction or conduction (see Appendix, A), had to pass through two floors of considerable thickness and two intervening rooms. No wire nor any other visible means of conveying the current was used.

CHAPTER III.

VIBRATIONS IN AIR AND IN THE ETHER.

IT is a common mistake to credit the vibrations in air with doing the work done by vibrations in the ether. It is quite true that there are instances in which electric currents promote vibrations in air, causing sound, and there are instruments made involving the adoption of this principle.

But we must draw a very wide line between the two kinds of vibrations in all experiments associated with wireless telegraphy.

One set of vibrations, those in air, has to do with, it may be, thousands of waves per second, but those in the ether are reckoned by hundreds of millions, hundreds, and even thousands, of billions per second.

We may easily see the difference in the behaviour of the two media, air and the ether.

Suppose in a thunder-storm, three miles away, we see a flash of lightning; the light-waves in the ether reach the eye practically at the same instant that the flash actually occurred, because, if light be capable of travelling round this earth about eight times in one second, it would not take the sixty-thousandth part of a second to travel three miles. But what about the noise occasioned in the air by the electrical discharge?

Noise or sound has to do with the other medium, namely, air, and its reception is accomplished by the ear, not the eye.

The waves in the air do not travel at the same pace. The average speed may be about 1,150 feet a second, according to the temperature of the air. This would mean about fourteen seconds for the thunder to make itself known to us. The electric current in fourteen seconds would have gone round the earth more than a hundred times.

With a moment's thought it will be seen that time may be left out of the question in speaking of wireless telegraphy, which has to do with the ether, and not air; ethereal waves, not sound waves.

Another illustration may not be amiss:

A skeleton clock, with hammer and bell visible, and wound up to keep ringing for a considerable time, is placed under the glass receiver of an air-pump. As the air is pumped out the sound gradually gets weaker; we see the hammer striking the bell, but the sound is almost nil, as the exhaustion approaches a vacuum.

You cannot have sound if there be an absence of air, for there is nothing to set into vibration by the vibrations of the bell, which has been set into vibration by the action of the hammer.

But there is a medium present which cannot convey sound-waves, that is the ether. It is there, otherwise I should not be able to see the clock. This is made visible to me by the light which reaches my eye. The light waves do not require the medium air, but ether.

In the accompanying scales of vibrations we see the great line of demarcation between waves in air and those in the ether. First, we note the ordinary range of human vocal powers, including those of professional singers, and observe the wonderful range of the powers of human hearing. But even hearing power has its limits, for there may be sounds with vibrations exceeding forty thousand per second, which are too shrill for our ears. This may be an advantage,

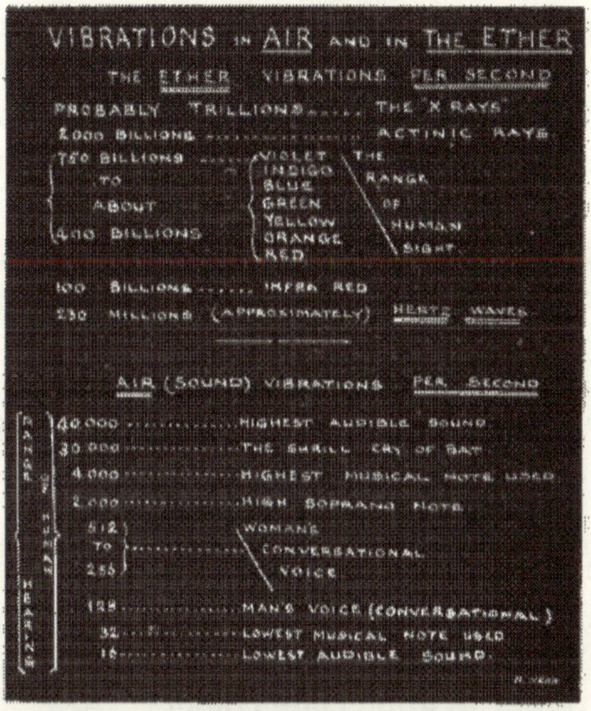

FIG. I.—SCALE OF VIBRATIONS.

for if the insect tribes be capable of producing still higher sounds, it is just as well we cannot hear them.

Some entomologists think that insects

have some such power of calling to each other, and, if so, it may account for the systematic way in which they meet in the summer months.

That there are higher sounds which we cannot detect is proved by Galton's whistle and fine gas flames, which are made to quiver by the sound waves. The whistle, as we screw it tighter and tighter, produces notes higher and higher, until at last we fail to hear any sound, but the flames still quiver in response to notes of the whistle which are inaudible to us. As one writer puts it, ' the flames hear what we cannot hear.'

To proceed now to the other set of vibrations, those in the ether, we are struck with the limited range of the human eye. It reaches from four hundred billions per second (red), to seven hundred billions (violet). But what a wonderful power, the power to see, the power to appreciate colour !

Next to the right use of the mind comes in order this power to see. The ether is the medium, light the agent, and the eye is the receiving instrument.

The adaptability of the eye to light is a marvel in itself.

But even if the eye be limited in its range, what does it matter? The mind of man applies itself, and other rays, invisible to his eye, are discovered and utilized in ways that are beneficial to mankind.

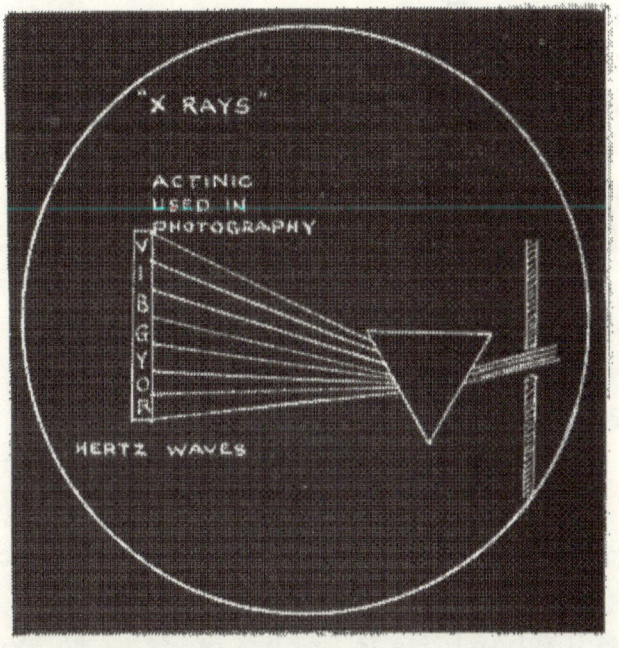

FIG. 2.—ANALYSIS OF LIGHT.

The photographer of modern times, taking hints from Daguerre, Reade, Fox Talbot, Archer, and others, calls in the invisible actinic rays above the violet (see Fig. 2), and these invisible rays make visible pictures for us.

With other invisible rays we have the benefits accruing from Röntgen and Lenard's discoveries.

The Hertz vibrations, also invisible, because not sufficiently rapid to be visible, and occupying a position below the red of the visible spectrum, are likely to be beneficial in saving life when wireless telegraphy is fully accomplished.

Thus the mind, when properly applied, is capable of discovering in the beam of light blessings that have lain dormant probably throughout all previous time. And, as already stated, we have not found out yet all that may be known about, or all that may be contained in, a beam of light. For a certainty, we have not found out much about the ether.

But what little is known leads us to expect great things in the near future.

By it we may yet understand a great deal more of the mysteries of gravitation and cohesion, and obtain clear perceptions of the causes of phenomena that are now very puzzling to us.

The ether conveys energy from the sun in the form of waves.

These waves vary in length. To one set we give the name of *light*, and the eye is adapted for the appreciation of those waves.

The surface of the body appreciates or feels other waves, to which we give the name of *heat*.

Other waves are detected by delicate instruments, and to them we give the names of *electricity* and *magnetism*.

As the eye receives light, so Lord Kelvin says the delicate coherer of Branly is an 'electric eye,' in that it is sensitive to electric waves of the Hertzian series.

We speak nowadays of energy taking on the form of waves, and though electric currents are said to 'flow' along wires, the expression is hardly accurate enough (see Appendix, B).

Although there is such a wide difference in rapidity between waves in *air* and those in the *ether*, there is a certain parallelism in their requirements and their behaviour; so much so that for purposes of practical demonstration we use experiments with waves of air to illustrate those in the ether;

hence the same terms are applied in each case.

This brings us to the law of sympathy, or 'syntony,' to understand which, as it is applied to wireless telegraphy, we must see its bearing upon musical instruments, and in fact upon anything that can be made to vibrate.

It is well known to musicians that if a violin and a piano be in the same room, and if *they are tuned to each other*, as if about to be used in a duet, a note sounded on the violin will find a response in the piano, if the dampers be raised from the strings by putting down the pedal. It is useless to try to hear any result without previously tuning the violin.

A blind gentleman skilled as a pianoforte tuner has stated that when in a lady's drawing-room tuning a piano something in the room responds whenever he touches a certain note. This is disconcerting to him, and he is obliged to search until he finds the offending article. Generally it is a vase on the mantelpiece. He reverses it, or so alters its position that it cannot respond.

The tuning goes on, and he is again inter-

rupted. There is something else respond-
ing when he touches another note. This he
finds is over his head. It is a screw used
for holding the globe of the gas pendant in
position that has not been secured sufficiently, and is just ready to vibrate in

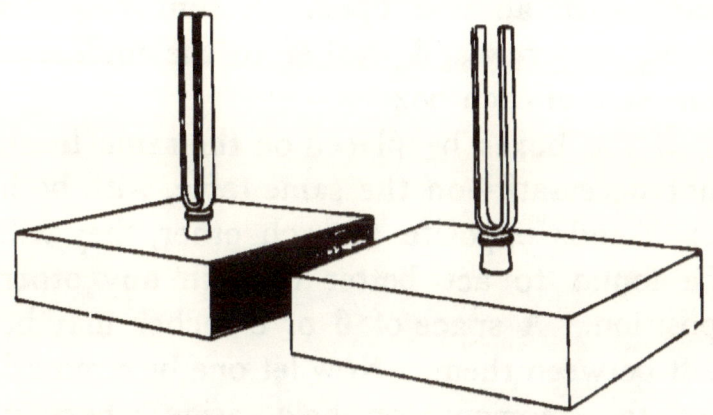

FIG. 3.—TUNING-FORKS TO SHOW SYNTONY.

sympathy with the vibrations of one certain
note.

There is this underlying law of sympathy
with regard to air vibrations that has to be
taken into account in many ways, as it
shows itself very often when not required
or expected.

To illustrate in a practical manner the

sympathy existing between two tuning-forks of the same pitch, we resort to the following experiment : In Fig. 3 two large forks, each made to give out 128 vibrations a second, are represented. They are about 1 foot in height. Two resonance boxes, equivalent to the sound-board of a piano, are shown, each with an end open. A tuning-fork is fixed by screw and washer to the middle of one side of each box.

If the boxes be placed on the same level, not necessarily on the same table, with both open ends opposite to each other, they will be found to act better than in any other position. A space of 6 or 8 inches may be left between them. Now let one be removed for the moment, or held aside where it cannot respond. The bow—that of a bass fiddle will do—should be drawn across the top of the two prongs of the fork on the table, when a volume of sound will be heard throughout even a large building. It should then be ' damped '—that is, the prongs should be held so as to stop the sound. Now let the other be placed in the position where it stood before removal. The two are now in a line with each other. If the bow be again drawn across the first tuning-fork as before, and after a few moments 'damped' again, it will cease to give out any sound; but the second tuning-fork, not touched at all by the bow, has taken up the vibrations from number one, and, with equal intensity, gives out the note first started in the other.

In much the same way there appears to be a good deal of sympathy between two wires of the same calibre and insulation (see Appendix, C) when they are placed in a *parallel* position.

In all Hertzian wave experiments, whether carried out by Signor Marconi or Dr. Lodge, a system of 'tuning' must be resorted to in order to establish perfect unison between the receiving apparatus and the transmitter. Yet the tuning does not refer to 'sound' at all.

It exists in the Leyden jar experiments, which are described in a subsequent chapter. The very name which Dr. Lodge gives to those jars, 'syntonic,' implies a similarity of tuning.

In Signor Marconi's receiving apparatus the two wings or 'capacities,' as noticed in another chapter, are used not merely as conductors of waves or currents to the little coherer, but as tuning accessories. Without these the receiver would not respond so readily or so accurately, if at all.

The sympathy cannot exist between the two main parts of the apparatus when removed far from each other unless one be tuned to the other.

So important does this tuning appear to be that the privacy of messages sent and received by wireless telegraphy may be secured by means of it.

Both Signor Marconi and Dr. Lodge are doing their best in their respective departments to establish a means by which one message sent through space may be differentiated from another, so that messages sent out may reach those for whom they are intended and no one else.

Mr. Swan, F.R.S., referring to this particular part of Dr. Lodge's work, says: 'Obviously it would be very inconvenient if messages sent through space were indifferently receivable by everyone who chose to play the part of eavesdropper.

'That condition of things would somewhat resemble that described in one of Hans

3

Christian Andersen's stories, where the fumes coming from a pipkin revealed to everyone who chose to smell them what each particular person was having for dinner. It was not very desirable that that kind of curiosity should be gratified in connection with telegraphy, and it seemed to him that the uses of telegraphy through space would be very much limited if this sort of thing could not be prevented. Professor Lodge's line of experiment, however, seemed to tend in that direction, and to show the means of confining a message to the person intended to receive it.'

If the whole accomplishment of this ideal in telegraphy without wires should be found to rest upon the natural law of sympathy, it follows that the system of tuning must be carried out to a nicety.

CHAPTER IV.

JAMES BOWMAN LINDSAY.

Born, 1799; Died, 1862.

In giving an account of any discovery or invention, it is but right to give honour to whom honour is due. Therefore we cannot refrain from paying a tribute to the memory of the man who, in these islands at least, was the first to suggest a method of signalling across space without intervening wires. Not only did Lindsay suggest, he also carried out successful experiments in proof of his theories.

It would not be easy to name a greater genius than James Bowman Lindsay, nor a more noble character. Certainly few, if any, have accomplished so much in a lifetime of penury. All his life long he must have pinched himself to the utmost limits in order to purchase materials for his numerous

experiments.   He worked, and worked alone,
on the borders of starvation.   He had no
house ; only one room could he afford, but
that one room had in it more than any palace

PORTRAIT OF JAMES BOWMAN LINDSAY.
By kind permission of Mr. W. M. Ogilvie, Royal Bank, Lochee, Dundee.

in the kingdom.   It was lit up by an electric
light of his own installation in the year 1835 !
It is difficult to realize that sixty-three years
ago a room could have been so illuminated.

That same room was famous for other reasons. It was here he wrote several of his works, and that portion of his dictionary in fifty different languages, which, in his own handwriting, is in a glass-case in the Dundee Museum.

When I saw that manuscript, and the vision of his life-struggle passed in a moment through my mind, I removed my hat in reverence for the memory of that poor but rich linguist of Dundee. A short account of what he accomplished under enormous difficulties will be acceptable now that telegraphy without wires is occupying so much attention.

James Bowman Lindsay was born in 1799, at Carmylie, and was taught weaving, and from earliest youth he endeavoured to educate himself. In 1821 he entered St. Andrews as a student, working at his trade during the college recess. In 1829 he was appointed lecturer and teacher at the Watt Institution, Dundee.

After finishing his Arts course, he became a divinity student, but never took license.

In 1841 he was appointed teacher in Dundee Prison at a salary of fifty pounds.

He was a diligent student of science, made many discoveries, and published many works.

It is possible that he was the first to use the electric light. Not only did he succeed in lighting his one room in 1835, but he publicly exhibited an electric lamp that same year in Dundee.

In 1845 he suggested the possibility of extending the electric telegraph to America.

In 1853 he maintained that it was possible to establish electrical communication through water without connecting wires.

In 1854 he patented this invention. That same year he conducted experiments in London and at Portsmouth, where he successfully telegraphed without wires across water 500 yards wide.

In 1859 he telegraphed in this manner across the River Tay at Glencarse, where it is about half a mile wide, and read a paper on this subject before the British Association assembled at Aberdeen. In presence of the members he conducted experiments at the Aberdeen Docks, which successfully proved the correctness of his theories.

During the last two or three years of his life he had a pension of one hundred pounds, granted by the Queen on the recommendation of Lord Derby.

In his experiments (see Appendix, D), Lindsay was very successful, but it is a matter of doubt whether his suggestion as to signalling to America would have met with success, even if facilities had been granted for the trial. At the same time, it is hardly fair to condemn the suggestion in the absence of the man. The genius of the mind that could invent wireless telegraphy, and the indomitable energy that could encounter the labour of writing a dictionary in fifty languages, cannot easily be set aside or even deemed in error: we do not know all the resources that were behind the suggestion.

That his was a prescient mind also will be seen in the remarkable words inserted in the advertisement announcing the opening of his science classes, which appeared in the *Dundee Advertiser* of April 11, 1834:

'Houses and towns will in a short time be lighted by electricity instead of gas, and heated by it instead of coals, and machinery will be wrought by it instead of steam, all at a trifling expense.'

Fancy all this foretold sixty-four years ago!

Some of Lindsay's experiments were made in the presence of a turnkey of the gaol and a young friend who is now a venerable old gentleman, highly respected in Dundee.

This friend, Mr. Loudon, senr., tells me that Lindsay would station them on one side of the Tay, requesting them to watch carefully the needle he had placed in position, and to note how it moved. He would then insert his plates in the water on their side of the river, and, crossing over to the opposite side, would complete his arrangements. His battery, containing twenty-four Bunsen cells, would be set in action. Later on Lindsay would return, and question them eagerly as to the behaviour of the needle. The different movements to the right and left would be noted, and, on comparing them with the messages he had sent from the other side he was perfectly satisfied he had accomplished telegraphy without continuous wires. He would then return to his little room as happy as possible.

It is comforting to know that the studious Lindsay, with all his poverty, got some

satisfaction and joy in this life. Nowadays
we can purchase batteries and wire and all
electrical appliances at a moderate cost, and
every detail ready for immediate service. It
was not so in Lindsay's day. He had to
make up his cells, and possibly to coat and
otherwise insulate his wires. Mr. Loudon
knows that Lindsay had to make his own
intensity coils, and he tells me that one
particular coil used in his earlier experiments
was $4\frac{1}{2}$ feet long, and contained 5 miles of
wire.

Think of it. Think of the cost, and
where the money was to come from. Think
of the patience of the man, with primitive
and limited appliances. The modern ex-
perimenter in electricity has no difficulties
compared with those combated by Lindsay.
Besides, in Lindsay's case more wire, more
acids, more metals, meant less food, less
new clothes, less comforts of all kinds.

Before looking into this remarkable room
of his—which, if now existing, would be a
sacred spot to the scientific mind of to-day
—let us refer once more to his practical
experiments.

One of the first demonstrations of his

theories took place at Earl Grey Dock, Dundee, in presence of many men of science.

He immersed separate plates in each side of the dock, and transmitted messages with ease and accuracy. Going to a wider expanse of water, he repeated his experiments across the Tay at Dundee and Woodhaven, where the river is nearly two miles wide !

These wonderful experiments were the chief topic of conversation at that time in Dundee, but nothing was done to give them a practical bearing.

This was not Lindsay's fault. The business of the philosopher is to find out the mysterious forces in Nature, and simply to indicate their application. It remains for others, who have the necessary capital and practical ability, to adopt the ideas and suggestions, and to shape them to a useful result.

The philosopher's part was done, and well done. If the term ' philosopher ' does not apply to Lindsay, it never applied to any man. How is it that there is no memoir of this man's life and doings ? Where is the statue to his memory in Dundee ? I have seen statues in large cities to the memory of men not worth remembering — men who rose to fame on other men's shoulders; men who, in point of morality and genuine worth, were not fit to enter his little room, or to polish poor old Lindsay's boots.

But if Lindsay's wishes could be ascertained, he would ask for no statue. It is more than probable he would suggest a large lecture hall, to which the public could have free access to hear unfolded the grand mysteries of Nature, to the study of which his own life was devoted.

Doubtless the time will come when the people of Dundee will do their duty, and honour themselves by honouring the memory of Lindsay. Few men who ever saw light on these islands of ours have so completely ignored self on behalf of science, or with so much earnestness of purpose have struggled against poverty and accomplished so much; for it must be borne in mind that ' Lindsay's work as a linguist was as remarkable as his scientific discoveries, and would have made a reputation for him rivalling that of Mezzofanti.'

But we have not yet peeped into those famous rooms of his. Sir W. C. Leng

writes: ' James Lindsay, the learned Carmylie weaver, died while I was in Dundee,
1862. His rooms (two rooms now that he had
a pension), on a flat near the harbour, were
walled round with books, and had stacks
of books on the floor. His great work in
manuscript — a dictionary of twenty-six*
languages, lay arrested by death on the
table.   '

' He had got a volume of ponderous bulk,
and had ruled it in narrow lines across and
across, so as to allow of the equivalent of
each word being written in many languages,
in a small hand, on the same line as the
original.

' Very pathetic was the testimony borne
by that book to the old man's ambition to
leave something monumental behind him,
and the manner in which his hand had been
stopped in the midst of his labours.'

* It was a dictionary of fifty languages.—R. K.

CHAPTER V.

IT will now be necessary to notice two methods of employing the good services of the ether.

One is brought about by *dynamic*, the other by *static* electricity.

By *dynamic* electricity we understand that which is said to 'flow,' guided by a conductor.

By *static* electricity, that which simply charges a conductor or the surface of any material capable of receiving it.

Though the terms 'flowing' and 'current,' as applied to electricity, may not be strictly scientific, it is not easy to avoid using them.

Mr. Preece has employed *dynamic* electricity in all his great experiments, and he terms the particular form of energy he employs 'electro-magnetic waves of low frequency.'

The late Dr. Hertz used *static* electricity, his waves being of high frequency, probably 230 millions a second. Unfortunately for science, this great man died in 1894, at the early age of thirty-seven. But he achieved great things, so much so that all workers since his day in this particular department of physics gladly recognise in him the originator of this line of thought and research.

Dr. Lodge pays a high tribute to the great industry of Hertz, notwithstanding the brief span of his life : ' The front rank of scientific workers is weaker by his death . . . yet he did not go till he had effected an achievement which will hand his name down to posterity as the founder of an epoch in experimental physics.'

Mr. Preece's experiments with dynamic electricity owe their origin to Faraday's great discovery of *induction.*

Faraday commenced his researches in 1831.

His investigations extended over a wide field of electrical knowledge, and he succeeded in establishing certain laws of electromagnetism, and elucidated many phases of electric phenomena which were little understood previous to his time. Faraday's three volumes, entitled ' Experimental Researches in Electricity,' form the basis of most of the great work accomplished in recent times in this branch of science.

By *induction* (see Appendix, A) we understand that a current of electricity passing in one wire sets up a *magnetic* field around that wire, so that electricity and magnetism are inseparables. They seem to be married to each other, and when electricity is invited out, the invitation might as well be extended to magnetism, as it is sure to be present whether invited or not.

Now, any other parallel wire within that magnetic field has a current of electricity *induced* in it whenever that field varies from the steady state, or either wire varies from the steady state.

From Faraday's time until the invention of the Bell telephone in 1877, everyone with a knowledge of electricity was aware of this law, but not of its limits.

Owing to the coarseness of the telegraph apparatus employed, inductive disturbances up to this time were unimportant, but on the introduction of the telephone—by far the most delicate detector of the presence of electrical vibrations known—induction asserted itself, and single-wire telephone circuits became unworkable within the magnetic field of any wire carrying varying currents.

One of the first prominent cases which came under Mr. Preece's notice was the Gray's Inn Road instance, where the telephone wires were disturbed on the tops of houses eighty feet high by the telegraphic work going on in the underground wires in the street parallel to those overhead.

This occurred early in 1884, and was got over by diverting the telegraph wires to a more distant route.

Mr. Preece, when lecturing before the Society of Arts in 1894, on 'Electric Signalling without Wires,' makes mention of a very exhaustive series of experiments made for him by Mr. A. W. Heaviside, the Superintending Engineer of the Northern Division of Postal Telegraphs. These experiments were of an exceedingly important character, and were made during the years 1885-1887 in the neighbourhood of Newcastle-on-Tyne.

At the time, they created a great deal of interest, and now that wireless telegraphy is commanding world-wide attention, and is no longer looked upon as a myth, it will be interesting to refer to the newspapers of ten years ago and see what was done in the way of experiment, either by *conduction* or by *induction*.

The following experiments, though probably not in the sequence in which they actually occurred, were reported in the *Newcastle Daily Chronicle*, the *Newcastle Daily Leader*, and other northern journals.

Among the experiments initiated in the north of England, one of the first was on the town moor at Newcastle.

The effects of parallel *induction* of Morse signalling were very marked, and this moor was found not big enough to get away from them.

The wires stretching from the Grand Stand on the moor to Blue House were experimented on. Strong intermittent currents were sent through one wire and the vibrations were distinctly heard by telephone in parallel wires laid out a quarter of a mile away.

Experiments were continued during the summer of 1886, culminating in signalling across a distance of forty miles between the east and the west coast main-lines of telegraph in the north of England. These were Morse signals; but it was subsequently shown that these results were not received across space, but through the medium of the great network of telegraphs that covered the railways and roads.

The next experiment in the north appears to have been that at Swarland Park, in August, 1886. Here it was conclusively proved that telephonic speech could be carried on inductively between two squares of insulated wire of one quarter mile sides, placed in parallel position, the distance between the nearest side of each square being also a quarter of a mile.

What had been proved at the outset was confirmed on this occasion, namely, that the signals fell off rapidly in intensity when the distance between the parallels exceeded the length of the parallels.

Even Professor Sylvanus Thompson, who is more careful than most authorities in drawing the line between *conduction*, *in-*

duction, and *wave* methods, admits that this experiment 'precludes all idea of earth conduction.'

Shortly after, another point was settled at Broomhill colliery, namely, that the electromagnetic waves were not interrupted by solid matter. This illustration appears to support previous statements respecting so-called solidity and the penetrative power of the ether. At the same time, the experiment may be the result of conduction purely.

An equilateral triangle of insulated wire three-quarters of a mile each side was arranged at the bottom of this colliery at a depth of 360 feet in a horizontal plane. A similar triangle was placed on the surface of the land, parallel to that below and exactly over it. Telephonic speech was heard and carried on between the telephones in circuit in each triangle.

Whether these experiments were the result of *conduction* or *induction*, the fact remains that the two triangles were separated by rocky materials of the earth 360 feet in thickness, and that distinct speech was conveyed from one triangle to the other and conversely! (See also Transactions of Electrical Engineers' Chicago Meeting, 1893.)

Mr. Preece also successfully sent messages across a distance of four and a half miles in the district between Gloucester and Bristol. He had great success in his experiments in the valley of the Mersey and at Porthcawl, in South Wales, where Mr. Gavey, the Superintending Engineer of the South Wales district, with squares of insulated wire 1,200 yards long laid side by side at various distances apart, and smaller squares above them, confirmed the fact that the results were due to electro-magnetic induction.

In the Bristol Channel also Mr. Preece succeeded in sending messages from Penarth to the Island of Flat Holm, a distance of three miles, and he obtained evidences of sound from Penarth to Steep Holm, a distance of over five miles (see Fig. 4).

The instances quoted of Mr. Preece's successes do not represent a tithe of what he has accomplished by laying hold of and applying Faraday's laws of induction to an extent that Faraday never even surmised. His lecture delivered before the Society of

Arts in 1894, and carried up to date as
delivered before the Royal Institution,
June 4, 1897, shows that as regards the
study of Nature's mysterious forces he is
an interpreter of high order.

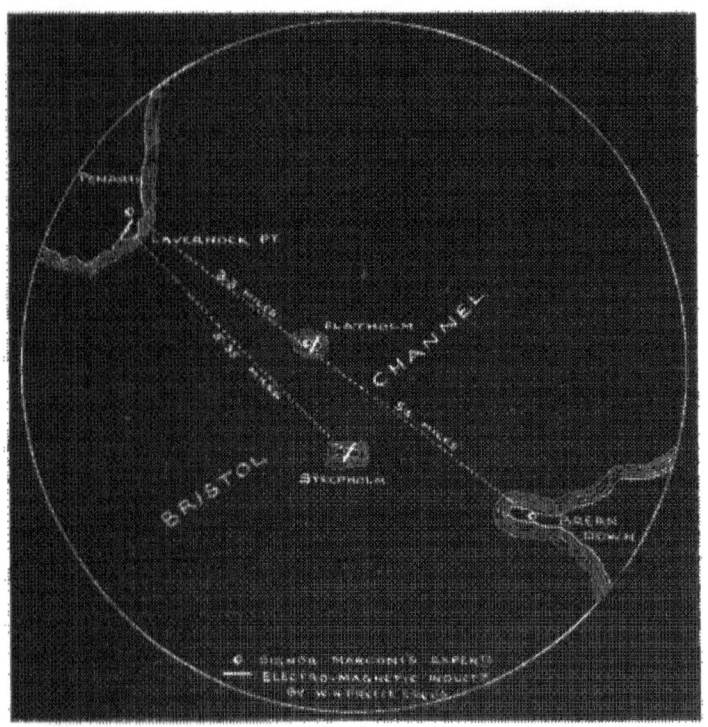

Fig. 4.—Part of Bristol Channel.

No man since Faraday's time has opened
up to us so much of the possibilities of the
laws of induction, or has shown more energy

combined with patience in carrying out experiments on such a gigantic scale.

If Mr. Preece does not succeed in beating all records as regards distance signalling through space without wires, the fault will not rest at his door.   In any case his work will make matters easy for those who come after him.

It should be understood here that James Bowman Lindsay's experiments were based on principles of *conduction*, water forming the conducting medium, whereas Mr. Preece has directed his attention to the properties of the ether of space and to the laws of *induction*.

In his lecture of 1894, already referred to, Mr. Preece raised curiosity and scientific thirst to the highest pitch as step by step he related his experiences in wireless telegraphy.   The concluding passage of that lecture I quote in full, because of its novelty of thought, and because it gives a more scientific foundation for speculation as to inter-planetary signalling than anything previously put forward :

‘ Although this short paper is confined to a description of a simple practical system of communicating across terrestrial space, one cannot help speculating as to what may occur through planetary space.

'Strange mysterious sounds are heard all along telephone lines when the earth is used as a return, especially in the calm stillness of night. Earth currents are found in telegraph circuits, and the Aurora Borealis lights up our northern sky when the sun's photosphere is disturbed by spots.

'The sun's surface must at such times be violently disturbed by electrical storms, and if oscillations are set up and radiated through space, in sympathy with those required to affect telephones, it is not a wild dream to say that we may hear on this earth a thunderstorm in the sun. If any of the planets be populated with beings like ourselves, having the gift of language and the knowledge to adapt the great forces of nature to their wants, then if they could oscillate immense stores of electrical energy to and fro in telegraphic order, it would be possible for us to hold commune by telephone with the people of Mars.'

For the purposes of demonstrating his theories on a small scale, so that audiences

may understand matters, Mr. Preece adopts
the following methods: In Fig. 5, A is an
accumulator, or other supply of electricity;
B is a Morse key; c is a buzzer as used in
field telegraphy; D is a roll of insulated
wire, say, of 150 yards. All these are connected, so that when the button of the
Morse key is pressed, these instruments and
wires are placed in circuit.   E is another roll
of wire similar to the other in every respect,
but not connected with it.   This roll of wire
is attached to a telephone trumpet F.   As
the button of B is pressed, the current passes
through the wire at D, and so influences the
buzzer that the whole audience can distinctly
hear its vibrations.

It will now be advisable to remove the
buzzer to another room with its wire con-
nections still intact, or to wrap it up so as
to deaden its noise.   When completely
subdued, so that the audience cannot hear it,
the telephone trumpet should be held towards
the audience, while with the other hand the
roll of wire E should be held over the other
roll as in the illustration.   The wire E will
catch up the energy from D by induction,
and the effect will be heard by the audience.

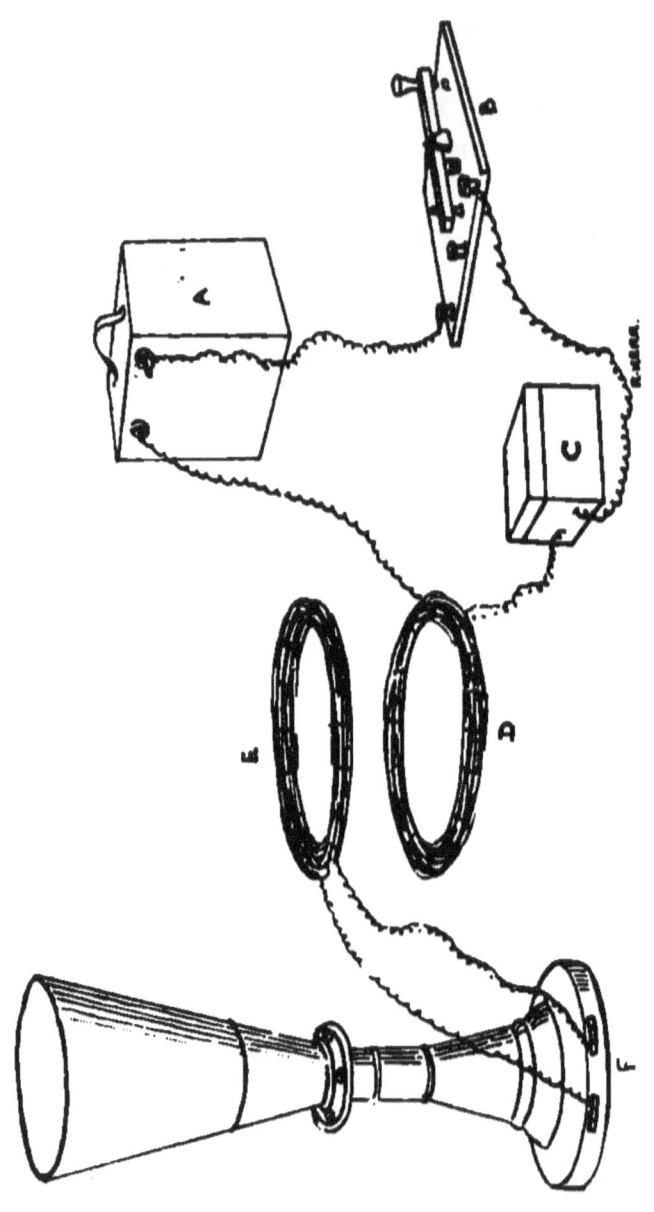

FIG. 5.—MR. PREECE'S INDUCTION METHOD.

The telephone will of course make the most of the current that reaches it, so that an audience of over three thousand people can distinctly hear the sound. The more the vibrations of the buzzer are suppressed or disguised, the greater will be the telephonic effect.

The nearer E is brought to D, the louder will be the sound produced.

Now by turning up the roll E in a vertical plane over D the sound will be lost, thus proving that parallelism must be observed.

In all Mr. Preece's experiments he has found it necessary to place the wire along which the current is sent, parallel to the wire in the distance that is to receive the energy and record the message.

There is so far no device so ingenious or so effective by way of bringing home to the capacity of non-scientific audiences the principles of *induction* as this.

Manifestly it is a disadvantage in several ways to have the two rolls of wire in such a small compass.

If each wire could be extended in a straight line for 150 yards, they might be placed parallel to each other 150 yards apart.   Then the message sent along number one and recorded at its other end, would at the same instant be taken up by number two and recorded at the corresponding end.

The lines would then occupy two sides of a square, the other two sides being absent, to use an Irishism.

In this way we see that if messages are to be conveyed over a channel of water three miles wide, there must be a three-mile length of wire on each side of the channel to obtain the maximum effect.

The square is there, but only two parallel and tangible sides of it are employed.

It would appear that apart from the Freemasons' mysterious use of squares and equilateral triangles, certain forces of Nature seem also to favour these Euclidian figures, as shown by the previous experiments both above and underground.

As no ordinary hall could admit of extended wires for this experiment, it is obvious the wires must be coiled up in a compact form. The effect is considerably reduced, but for all that the result is most conclusive and telling. No one need go away in ignorance of the law of induction.

Few laws are more wonderful. A current is sent along a wire (to use an expression that may not be reckoned scientific); immediately a magnetic field is set up all around that wire extending for, not yards merely, but for miles, and that influence sets up another current in a similar wire miles away, if it be parallel to the first wire.

The results are astounding. No one from Faraday's time to the present can tell where the magnetic field around a wire has its limits.

Sympathy on a tremendous scale seems to play an important part between the two wires, notwithstanding the great distance which may separate them. But this will be dealt with in another chapter.

We must now turn our attention to the experiments initiated by Hertz with Static Electricity.

CHAPTER VI.

HERTZIAN WAVES—EXPERIMENTS BY SIGNOR MARCONI.

THE original and classical experiments and lines of thought suggested by the lamented Hertz have been followed by many able men, notably by Dr. Oliver Lodge, Signor Marconi, Professor Nikola Tesla, Professor Auguste Righi, M. Edouard Branly, and others.

Lord Rayleigh, in his lecture on ' Rapid Electrical Vibrations,' delivered at the Royal Institution, April 10, 1897, states that ' ob- servation of small sparks was the method employed by Hertz himself, who found that when an all but complete metal hoop was within the influence of strong electrical dis- turbance, minute sparks occurred at the gap.'

Dr. Oliver Lodge also speaks of this in his book on ' The Work of Hertz.'

This extremely simple occurrence observed by a thinker may be looked upon as an instance of the first transmitter and first receiver among scientific appliances.

The electric machine causing the ' electrical disturbance ' was the transmitter or oscillator, while the receiver consisted of the ' all but complete metal hoop.'

From these observations great results have followed.

It is apparent that all experiments in wireless telegraphy involving the employment of *static* electricity must necessitate the use of two pieces of apparatus—that which sends out the waves, and that which receives some of those waves.

Bearing directly upon the experiment of Hertz just referred to is another by Dr. Oliver Lodge, which he performed in presence of Lord Kelvin, in 1890.

In Fig. 6 there are two Leyden jars of the ordinary type, A and B, made as much alike as possible.   The jar B is connected with an electrical machine.   The jar A, although apparently the exact counterpart of B, requires a slider E, which can be adjusted until perfect ' syntony ' exists between these two jars. The jar A has no tangible connection with the other jar nor with the machine. A strip of tinfoil is placed over the edge at D, which is connected with the tinfoil forming the inside lining of A, and approaches the tinfoil on the outside at D, but does not touch it. Now arrange the circuits so that their planes are in parallel position and their

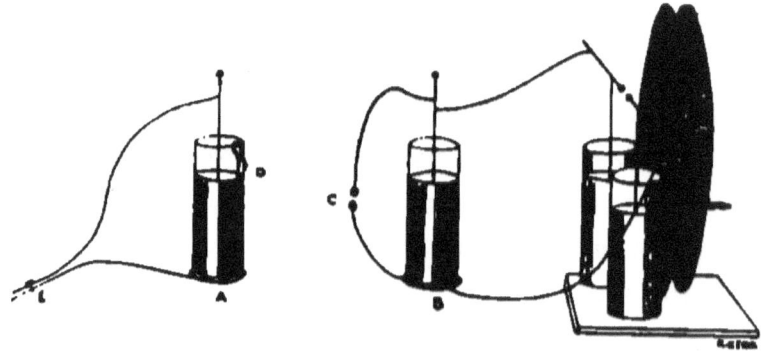

FIG. 6.—DR. LODGE'S SYNTONIC JARS.

distance not more than two or three times their diameter.

If the machine be set at work, the jar B will be discharged at C, and the instant this discharge takes place there will be a bright spark observable at D, the result of sympathy or 'syntony.'

This brings us directly to the consideration

of apparatus and phenomena of still higher order, but based on the foregoing and similar experiments.

A *transmitter* as it is now used by Marconi is comprised of three parts. Roughly speaking, the *receiver* consists of four parts.

For the transmitter we require a battery, or a charged accumulator, an intensity coil, and for ordinary use a stand supporting two or four brass knobs, which must be solid throughout, and very highly polished.

For long distances, when four knobs are used, the two inner ones are half immersed in vaseline oil. This gives greater energy to the spark, and hence to waves.

A few further particulars as to the transmitter are necessary.

The accumulator should consist of four cells, thus forming an eight-volt battery.

Let this be represented at A (Fig. 7). The wires leading from the accumulator are joined to the intensity coil B, of 6-inch sparking power in air. Less than this, even a coil giving a spark of 2½ inches, will do for lecture demonstrations or ordinary experiments. Wires are led away from the coil to the brass knobs c.

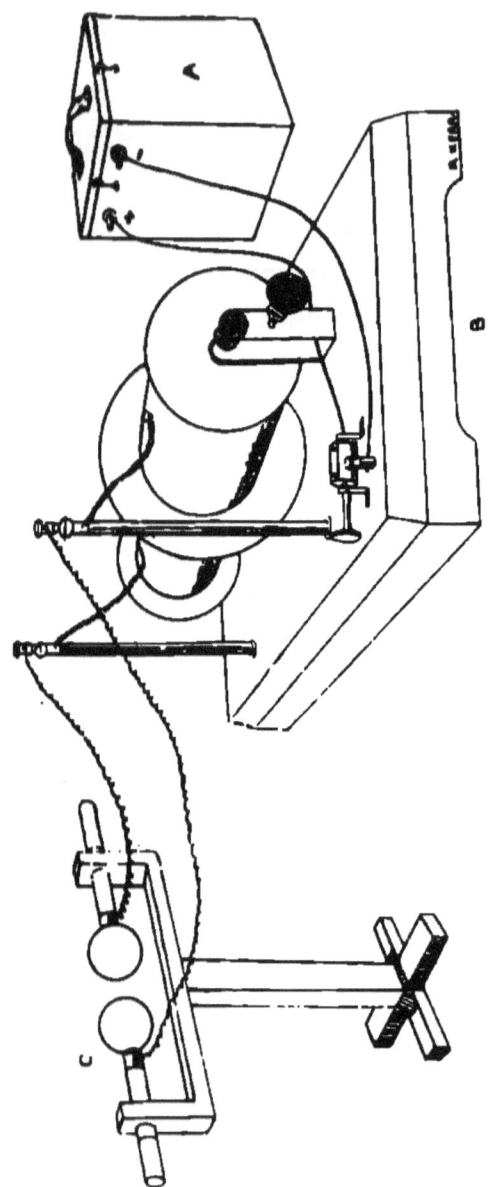

FIG. 7.—TRANSMITTING APPARATUS.

When the current is switched on, sparks rapidly pass between the knobs. These give rise to Hertzian waves, which travel out in all directions. Imagine the spark gap located at the centre of a hollow sphere, the waves would radiate from the centre to every point on the periphery.

As already stated, the waves are about 230 millions a second, and travel with a velocity of 186,400 miles a second, nearly eight times round our globe in that short period of time.

The distance between the brass knobs can easily be regulated at will to produce the best results. Care should, however, be taken to switch off the current before handling the knobs.

The transmitter is now complete. We are ready to send energy away into space through the walls and roofs of buildings to remote distances, and we are half-way towards a partial understanding of one of the most wonderful discoveries ever revealed by the mind of man.

I say 'a partial understanding' because I do not think that its chief exponents can grasp more than half of what occurs in the phenomena involved.

It is comparatively easy to work out certain experiments, but it is quite another matter to discover the underlying principles.

We must now make ourselves familiar

PORTRAIT OF SIGNOR MARCONI.
(From *McClure's Magazine*, by kind permission.)

with the wonderful apparatus which receives a share of the energy we send from the transmitter.

5—2

We might have the most perfect means of sending out the waves into space, but, without a sensitive device for arresting some of that energy, and recording the message wrapped up in that energy, we should have no results whatever.

There are many delicate instruments for effecting this object, but before noticing the most important part of any of these receiving instruments, namely, the coherer, we ought to look at the receiving apparatus in the concrete form ; details will be more helpful later.

Suppose, in order to see the construction of the receiving apparatus, we dismiss from our minds for the present the details of the transmitter, and imagine it five miles away.

The receiver consists of the delicate coherer attached to the glass rod AB in Fig. 8, a battery of twelve cells G, a separate cell F, wings C and D, and a post-office relay H (see Appendix, E). To this may be attached an electric bell K, or a Morse inker.

Other parts are described in the Appendix. The several appliances comprising the receiver are connected together. The current is ready, as it were, for completion, but there

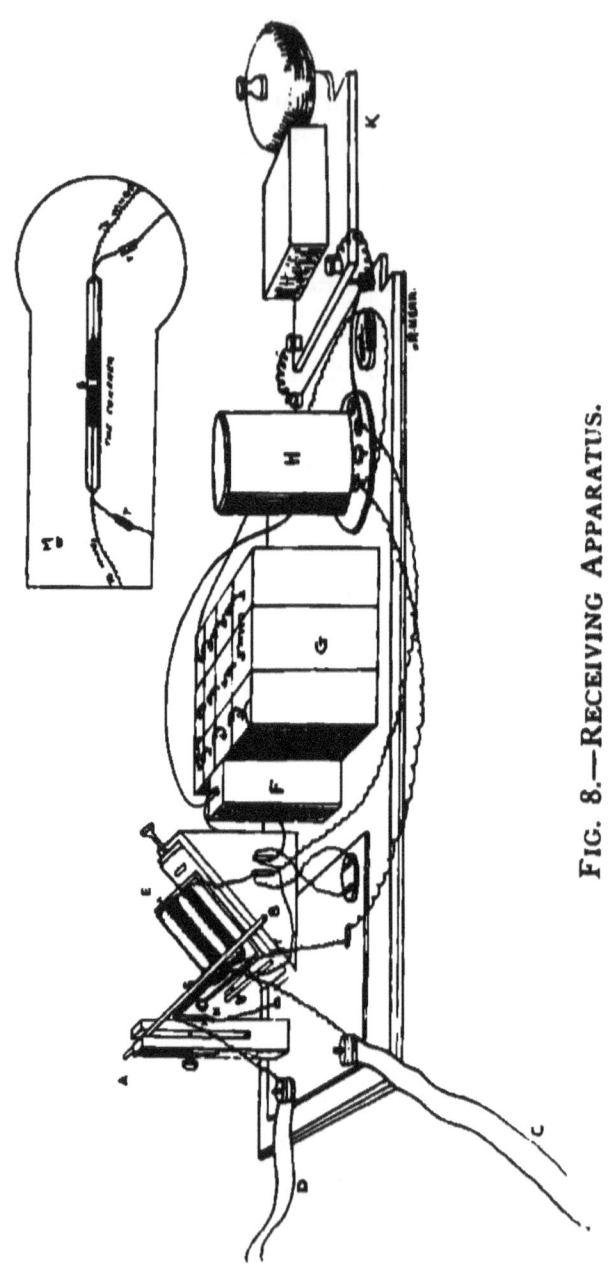

FIG. 8.—RECEIVING APPARATUS.

is a tiny gap in the coherer which must be bridged over ere the current actually exists.

Let us now consider the make of the coherer. We have selected Signor Marconi's, believing it to be the most sensitive yet con-structed. It will only be possible to give a very superficial description of it, and of the other parts of his receiving apparatus.

A tiny tube of glass, nearly 2 inches long, is provided. One wire enters one end of this tube; another wire enters the other end. These wires are connected with other parts of the receiving apparatus in such a manner that if the ends which pass into the glass tube were allowed to touch each other, the circuit with the battery, relay and bell would be completed, and the bell would ring. But the wires are kept apart to the extent of a space about the sixteenth of an inch.

The wires are terminated by blocks of silver, which closely fit into the tube, leaving the gap.

This glass tube is partially exhausted of air and sealed up, but not before a tiny pinch of metal filings is carefully inserted in part of the space left between the two silver ter-minals. The whole of the sixteenth of an

inch of space must not be filled for reasons
which will be given later. So long as these
filings remain loosely together in this space
they do not form a bridge for the current to
pass over from one silver block to the other.
But suppose the current to be switched on
at the transmitter, which we are imagining
to be five miles away from the receiver, it
travels on through the ether of space in all
directions, and part of the waves fall on the
coherer, causing the filings to cling together
or cohere. The bridge for the current is
thus made; the action at the coherer is
immediate, notwithstanding the distance of
five miles which separates both parts of the
apparatus, because of the amazing velocity
of electricity.

So soon as the filings become cohered the
bell rings, because the circuit is now completed.

One step more. The current when completed causes a small hammer N to tap the
coherer gently. This causes the filings to
separate, or decohere. There is room in the
space left for this separation of the filings. The
current is broken, and the bell ceases to ring.

Another current is switched on at the

remote transmitter, the filings are again instantly cohered, the current is again completed, the bell rings, the hammer taps the glass coherer, the filings are again decohered, the current is broken, and so on.

Thus we have power to make a current, or to complete it in the distance, and to break it as often as we like.

Hence we have means of sending definite messages over great distances.

Now, we ought to know something more about the filings (see Appendix, F). These consist of a mixture of silver and nickel filings, with the merest possible trace of quicksilver. Nickel lends itself splendidly to the reception of delicate currents in the ether, and is quickly decohered.

It will be easily seen that the same energy which causes a hammer to strike a bell will, if the bell be removed, be sufficient to cause a pen or pointed instrument to strike a strip of paper which passes under it by clockwork. Thus the paper will contain punctures or ink-dots corresponding with the number of times the current has been switched on at the transmitter, and with the same precision of intervals.

This is, in fact, the duty taken up by the Morse inker.

To return again to the little glass coherer. There is a plan whereby energy is caught and conveyed to the coherer. Two wings of copper stand out in line with the glass tube, and are joined by metal contact to the wires leading into the tube. These catch or inter-cept waves in the ether, and convey the current to the filings (see Appendix, G).

In Fig. 8, attached to AB is the glass tube ; the dark lines in it are the silver blocks forming the terminals of the wires. s is the space, occupying part of which are the filings. E is the electro-magnetic ap-pliance for tapping the coherer. N is the knob of the tapper. M is a separate sketch of the coherer. TT are choking coils (see Appendix, H). C and D are the copper wings, which, as already stated, convey energy to the filings. It is chiefly by them that the ' tuning ' of the receiver to the transmitter is brought about.

Before the receiver can respond to the energy sent from the transmitter it must be in sympathy with or ' tuned ' to it. The lengths or superficial areas of these wings must be adapted by gradual trials to suit the coherer. And if this be done perfectly the receiver ought to respond only to such transmitters as are absolutely in tune with it.

In another chapter, where the law of

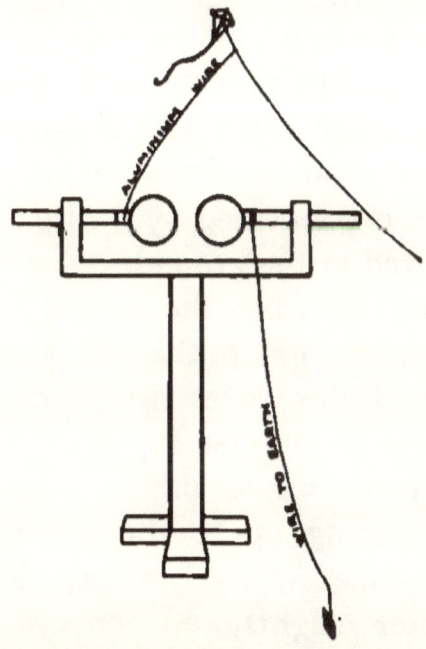

FIG. 9.—KNOBS AND KITE.

sympathy is described, this question of 'tuning' is again referred to.

So far we have given a superficial description of Signor Marconi's coherer, and the general requirements of his receiver. But for popular reading it will be unnecessary to explain in detail his system of resistances, his choking coils, and his relay. A fuller account of these will be found in the Appendix.

When Signor Marconi is signalling to places several miles off he attaches one end of an aluminium wire to one knob of his transmitter; the other end is fastened to a kite flying several feet from the ground. The other knob is connected by a wire to a metal plate inserted in the ground.

It follows that when the current is switched on, the wire attached to the kite is electrically charged, and the waves necessarily proceed from a great height (see Fig. 9).

The coherer has a similar arrangement; the wire leading up to the second kite in this case collects, as it were, the energy from greater heights, and conveys it to the filings in the coherer. The receiver must also have an earth connection (see Fig. 10).

He has sent messages across the Bristol Channel over a distance of nearly nine miles. Since then, when in Italy, he placed his transmitter in the fortress of San Bartolomeo and his receiver on board the ironclad *San*

Martino, removed twelve miles to sea. The
receiver was placed under the guns and in
the engine-room, surrounded by tons of
steel, yet the messages were recorded with
accuracy.

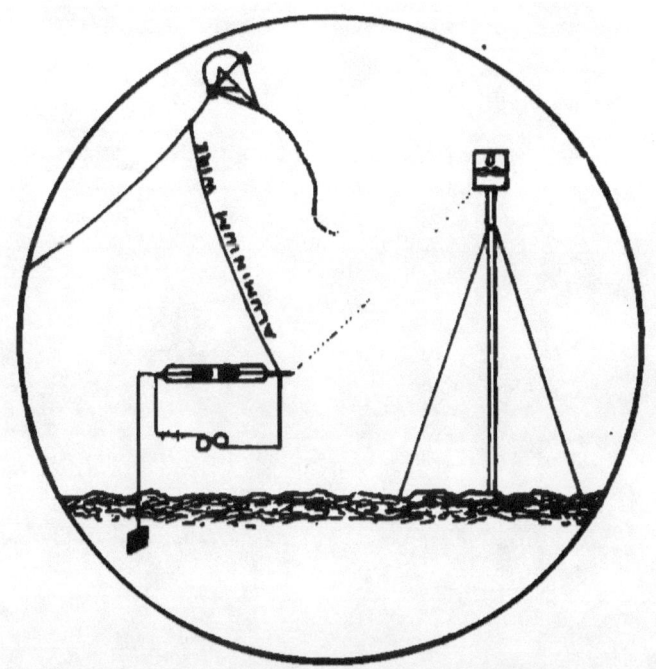

FIG. 10.—COHERER AND KITE.

Afterwards he successfully signalled over a
distance of seventeen and a half miles, from
the Needles Hotel, Alum Bay, to Swanage.
The transmitter was placed at the hotel,
while the receiver was on board the steamer.

Messages were received and recorded during the journey to Bournemouth and onwards to Swanage. Each dot on the ribbon in the Morse inker attached to the receiver corre-

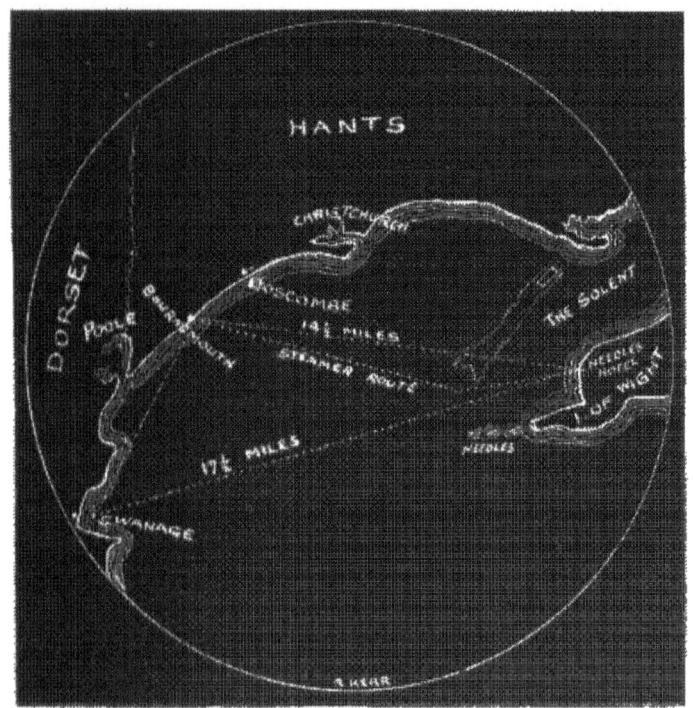

FIG. 11.—BOURNEMOUTH STATIONS.

sponded with the number of times the Morse key was pressed in the hotel in the Isle of Wight (see Fig. 11).

The weather was stormy—just the kind

required for testing the apparatus — but hardly appreciated by the gentlemen in charge of the receiver.

Successful signals have been sent from Salisbury to Bath, a distance, as the crow flies, of about thirty-four miles !

Recently, under Signor Marconi's supervision, a wireless telegraph-station has been established at Bournemouth on a practical basis. Messages are sent to and from Alum Bay, a distance of nearly fifteen miles.

But his system of telegraphy is not likely to be limited to peaceful purposes only, for if a lancer can hold a wire aloft and keep up conversation with his troop a long way off, we are likely to see it adopted for purposes of war.

Signor Marconi admits that he is engaged in the fitting up of apparatus for foreign Governments, therefore it does not need a great stretch of genius to surmise some of the ends to which his transmitters and receivers will be applied.

Contrary to the expectations of those who thought it impossible to send messages during foggy and stormy weather, the results are an improvement on those carried out when the air is clear and the sea smooth.

Successful signalling has been maintained from the ship during a storm and the station on the land.

In all efforts to save life along our coasts this will be a great gain, as instructions can be conveyed with ease from ship to shore and from shore to ship.

One of Signor Marconi's receivers was despatched from Bournemouth to Swanage. Evidently it was tuned to two transmitters, for it recorded messages from the station in Alum Bay and messages from Bournemouth.

Recently, in trying for results when the weather was stormy, the furthest distance they could get was to Swanage pier, nearly eighteen miles from the station in the Isle of Wight. The steamer carrying the receiver had a low mast of 55 feet, but with this height they were able to satisfactorily read signals from Alum Bay printed on tape, consequently they were not much surprised when they found they were able with a wire attached to a pole on top of a cliff to take in messages at Durlston Head, which is a little further.

The Wireless Telegraph and Signal Company are now putting up a permanent station at Durlston Castle. This will be the third on the south coast.

From the foregoing results we may expect to see stations established on each side of the Channel before many weeks, then between Holyhead and Dublin.

Lighthouses and lightships, as mentioned on another page, are sure to be included in Signor Marconi's list of early operations.

CHAPTER VII.

HERTZIAN WAVES—EXPERIMENTS BY DR. LODGE AND OTHERS.

IT does not appear to be necessary, for use over comparatively short distances, to follow out all the details just given with regard to the coherer and its accessories.

According to Professor Sylvanus Thompson, the Rev. F. Jarvis Smith, of Oxford, has for more than eighteen months maintained communication between his house and the Millard Engineering Laboratory, over a mile away, using a carbon powder coherer.

It does not, however, follow that such a coherer would do the work done by Signor Marconi's, *e.g.*, take in messages at Swanage, sent out from the Needles' Hotel, Isle of Wight.

Professor Thompson also states that Professor Slaby of Charlottenburg 'abandoned every one of the novelties introduced by Marconi, and fell back upon methods previously known. He used a simple Lodge-Branly coherer. . . . The elevated conductors were wires raised by means of hydrogen balloons to heights of nearly 1,000 feet. Signals were obtained at a distance of twenty-one kilometres, or over thirteen miles.'

In fairness to Signor Marconi the following extracts are made from Dr. Slaby's recent work : 'Marconi must have clearly added something new to the knowledge by which distances of a kilometre long were attained. I at once resolved to go to England, where the telegraph authorities were making greater experiments. . . . In the first instance, Marconi has devised for the process an ingenious apparatus, which, by the simplest means, attains certain results. He has thus first shown how, by connecting the apparatus with the earth on the one side, and by using long extended vertical wires on the other side, telegraphy was possible.'

These extracts go towards proving two points, at least, in Marconi's favour: first, that Dr. Slaby admits the originality of his work ; and secondly, that Dr. Slaby's success in signalling thirteen miles is attributable to Marconi's plan of using elevated wires. This appears to be so, for Dr. Slaby admits the novelty of using extended wires, and we find in Signor Marconi's specification of 1896 mention is made of balloons in the following words :

' Balloons can also be used instead of plates or poles, provided they carry up a plate, or are themselves made conductive by being covered with tinfoil. As the height to which they may be sent is great, the distance at which communication is possible becomes greatly multiplied. Kites may also be successfully employed if made conductive by means of tinfoil.'

We have, therefore, Dr. Slaby's experiments with a 1,000 feet of wire resulting in signals reaching a distance of over thirteen miles. And to contrast with this we have Signor Marconi experimenting with wires on a pole 150 feet high at Alum Bay, and sending messages to Swanage, nearly eighteen miles away ! If increased height of wire means even approximately proportional distance, Marconi will soon manage his sixty miles.

It should also be known that, previous to

these, his most successful works, Dr. Slaby spent a week in England with Marconi, and became acquainted with all his plans and methods.*

The *Electrical Review*, April 22, 1898, contains a description of an ingenious alteration made in connection with the Marconi coherer. It is called ' A Simplification of the Marconi Receiver.' The alteration is made by Dr. Rupp, of Stuttgart. It dispenses altogether with the electric tapper, and decoheres the filings by causing the tube to rotate on its axis. The leading-in wires of the coherer are mounted in bearings, and a small vulcanite pulley, with flanges, is fixed on one end of the tube. The tube is put in circuit by two small copper springs, which rub on the rotating axis. The strip of paper of the Morse instrument passes round the pulley between the flanges. The paper spool is slightly braked by a brass spring, in order to produce a uniform tension on the paper ribbon, and thereby a uniform rotation in the coherer tube. The diameter of the coherer tube must not be too small,

* Dr. Slaby was also present with Mr. Preece in his Bristol Channel experiments, and has repeated them before the Emperor of Germany at Berlin.

and the quantity of filings between the silver electrodes must be small enough to roll round the glass wall of the tube.

This is most ingenious, and apparently would answer all the purposes of laboratory or lecture-hall experiment, but its value in telegraphy depends upon the number of miles of space across which it can record messages with accuracy.

Dr. Oliver Lodge, in his famous lecture of 1894, showed several coherers consisting of glass tubes containing filings of different kinds.

He also suggested devices for tapping back filings to their normal condition.

Another of his suggestions is given in his own words : 'And by mounting an electric bell or other vibrator on the same board as a tube of filings, it is possible to arrange so that a feeble electric stimulus shall produce a feeble, steady effect, a stronger stimulus a stronger effect, and so on.'

So far as the Hertzian wave researches are concerned, the two great authorities are Signor Marconi and Dr. Oliver Lodge. But if results later on should prove that Signor Marconi is utilizing a new set of waves, say a set more penetrating than those of Hertz, we should have a case on all fours with that of the X-Ray discovery, where Lenard hit

PORTRAIT OF DR. OLIVER LODGE, F.R.S.
From a Photograph by Messrs. Elliott and Fry.

upon the cathode rays and Röntgen on those called after his name.

And this seems not at all unlikely.

Without any extraordinary battery-power, Signor Marconi seems to be working with waves that will penetrate anything, so much so that reflectors are useless.

Dr. Oliver Lodge, using apparatus apparently unlike that of Signor Marconi, may bring about results quite different, but of equal value to science.

It is more an advantage than otherwise that two keen minds should be engaged in work that is bound to be of unlimited benefit to the human race.

No stronger testimony to Dr. Lodge's powers can be found than that afforded by Professor Sylvanus Thompson in his lecture to the Society of Arts recently. Referring to the coherer, which is the most important part of the whole apparatus, he says : ' After the coherer has thus operated, it usually remains in the conductive state until subjected to some mechanical jar or shock. Lodge proposed to apply for this purpose a mechanical tapper worked either by clockwork or by a trembling electric mechanism. On several occasions, and notably at Oxford in 1894, he showed how such coherers could be used in transmitting telegraphic signals to a distance. He showed that they would work through solid walls.'

It is fully expected that at some of the gatherings of the learned societies this year Dr. Lodge will be in a position to show some entirely new phases of his work, and, among them, improved methods of distance signalling.

Recently, by introducing a few turns of wire between the spark gap and the wings, he made the transmitter send out longer trains of waves. This, it is said, facilitates the accurate tuning of the oscillations, and does away with the need of an oil receptacle for the brass knobs first used by Sarasin.

Many times at public meetings have questions like the following been asked: How will my receiver be guarded against messages coming from any and every quarter? How are the messages sent to me as matters of private business to be prevented from going to the receivers in other houses?

Dr. Lodge is not only endeavouring to secure all this, but much more. He is actually devising means that will make his transmitter a receiver, and his receiver a transmitter, so that each instrument will do double duty.

This in itself will place his appliances

ahead of everything he has done up to the present time.

It must not, however, be imagined that Signor Marconi is unmindful of the requirements of a perfectly private coherer. He has given his attention to this and to several other points that were considered insurmountable difficulties.

With regard to the similarity between the apparatus made and used by Dr. Lodge, and that used by Signor Marconi (if any similarity exists), it need not be considered in any way remarkable, inasmuch as both are working on the lines suggested by Hertz.

Marconi in Italy, intimately acquainted with Righi's work as well as with that of Branly, improves upon coherers that were filled up with filings, reduces the filings to a minimum—in fact, discards all previous filings—ascertains the kind of filings best suited for instantaneous *coherence* and *decoherence*, makes an ingenious contrivance not previously in operation which automatically taps the coherer at the exact moment necessary, and, to crown all his former work with the proof that he thoroughly understands what he is about, and a good deal more than some of his masters, he succeeds in telegraphing through space by waves to greater distances than any one has ever done in Europe. His experiments, too, cannot be attributed to earth currents, to induction, or to conduction, but solely to waves in the mysterious ether.

We should expect points of similarity to exist between the apparatus used by him and that used by Dr. Lodge. In fact, if there were none it would be remarkable, inasmuch as Dr. Lodge is also acquainted with everything done by Hertz, Branly, and Righi.

If an exhibition were held of each set of appliances, there would be found very little common to both.

Such an exhibition would be of great educational interest. It would settle all questions as to originality of ideas, and it would not fail to add lustre to both their names.

There is another name which must by no means be left out of our consideration in questions where electric currents of high frequency play an important part. Nikola Tesla, who a few years ago astonished the scientific world by taking through his body with impunity currents of electricity extending to hundreds of thousands of volts, and who is at the head of matters electrical in New York, is also at work upon wireless telegraphy. Some years ago he prophesied that not only messages, but electric power, would be sent without wires to Australia. It is generally believed that he has sent messages through twenty miles of ground without wires. And we are prepared to hear great things as the result of his genius. Realizing that the ether permeates all things so-called solid, it is said that he maintains the possibility of sending influences to the Antipodes through the earth (see Appendix, M). This is second in enterprise only to Mr. Preece's suggestions as to receiving messages from the inhabitants of Mars. In the 'Echoes of Science' paragraphs in the *Globe*, July 9, 1897, the following statement is made in reference to his twenty mile performance : 'In fact, by his own account he seems to have found the problem rather easy to solve ; but he will not vouchsafe any information as to how he does it until his work is complete. All we know is that he employs vibratory currents of electricity and waves,

or oscillations set up by them in the lumini-
ferous ether, which is coming to be not only,
as many believe, an inexhaustible store of
power, but a universal vehicle of power, a
kind of all-round omnipresent gearing.'

Nikola Tesla was born at Smiljan, a small
village on the Austrian border, and he is
forty years of age.

Evidently we may expect great things this
or next year. New discoveries of an astounding character, and improvements on recent
ones, are fairly well indicated.

CHAPTER VIII.

THE USES OF WIRELESS TELEGRAPHY.

THE advantages of a perfected system of signalling through space are very apparent.

For example, there were times in the Chitral campaign when a handful of men were in a fortress with limited provisions and limited ammunition. For days they were surrounded by thousands of the enemy. Their chief means of signalling could only be by the heliograph, which meant that they could only send out or receive signals during sunshine.

If they had had some such method as these under consideration, they could have sent messages by the transmitter which would have acted on the receiver in the camp miles away, and that again on the Morse inker.

The tape would pass along with its dots and intervals, and would contain all particulars as to how many men they had, how much ammunition, how much provisions, and how long they could hold out.

In fact, it would at all times be of immense value in all departments of the army and navy.

There would arise instances in naval engagements when it would at least be an auxiliary to the regulation methods of signalling from ship to ship.

Fogs, however dense, snowstorms, however severe, rain, even in torrents, or wind, however strong, would not militate in the least against the sending or receiving of messages. The ether permeates them all, and therefore weather does not come into consideration. It is doubtful even if a thunderstorm could in any way prevent such signalling.

The more one looks at the extraordinary difficulties wireless telegraphy surmounts, the more amazing it appears.

In the question of lighthouses it will be invaluable, not only as regards the keepers on duty and their additional connection with the mainland, but if the system is to enable one ship's company to signal to that of another ship, say, in mid ocean, then the lighthouse man can signal to the passing ship in all weathers, in darkness or in thick mists.

The sailor dreads a dense fog or a snow-storm more than a storm of wind. Therefore this new departure in science will be welcomed by all 'who go down to the sea in ships.' When foghorns, sirens, semaphores, bells, or even guns, are useless as signals, messages will be transmitted and received with precision at sea.

There are localities that are not easily connected by telegraph wires, where these systems will be available, such as islands separated by rapid currents where cables could not be laid or maintained without immense cost; also towns, separated by im-passable mountains, to the people of which wireless telegraphy would be a boon. The men leading monotonous but most useful lives away out in lightships would be in touch with the land. By them its advent would be hailed with delight.

There are numerous other ways in which wireless telegraphy could be utilized with advantages over even the wonderful system now established.

If Signor Marconi succeeds in signalling to Cherbourg—a distance of sixty miles from his Isle of Wight station—he will have no great difficulty in doubling that distance. By establishing signalling stations at intervals of a hundred miles, wireless telegraphy over large land areas would become practicable. This in itself would be wonderfully advantageous, and would appear to be attended with less expenditure than the present wire system. It is the opinion of more than one electrician that signalling to America and to the Cape will eventually follow.

It is difficult to realize that signalling without wires is an established fact already in full operation in the South of England. The Wireless Telegraph Company have two stations, between which signals and messages are passing all day long. One is located in a room at the Needles Hotel in Alum Bay, and the other is at Bournemouth, fourteen and a half miles away. The third station is in course of construction, near Swanage, at a point nearly eighteen miles from the Isle of Wight station.

On entering either the room at Bournemouth or that of Alum Bay, you expect to see a great display of ingenious and bulky machinery, with an amount of dynamos and the complex assemblage of heavy plant usually associated with electrical work. But nothing of the kind is to be seen.

One's ideas of great work are generally connected with pictures of big wheels and evidences of great noise. But here we find only a box that looks small enough to be carried in one hand, and the electric supply (a set of small accumulators), a six-inch intensity coil, a small Morse key, and wires leading to two small brass knobs highly polished.

This is all that is needed for the safe despatch of the messages.

But as this room is also a receiving as well as a sending out station, there must be something else for taking in and recording messages. Where can this be ?

It can be nothing else than that box about $2\frac{1}{2}$ feet long. When it is opened we see it contains a small Marconi receiver, comprising the coherer, relay, tapping apparatus, and a set of small cells forming a battery.

Near this box, and connected with it by

wires, is a Morse inker, with its paper ribbon passing out with its dots, spelling out messages now arriving from Alum Bay.

It is hardly fair to say this is all, because a wire is connected with the transmitter to a pole outside about 100 feet high. The receiver is also similarly connected. A few particulars as to the functions of this pole may be added here, although this accessory has been already referred to.

Wires in the form of netting are suspended from the top of the pole, and are well insulated, so that any energy conveyed to the wire netting from the adjacent room, or from the station away at Alum Bay, cannot travel down the pole to the earth, but only along the wire from and to the room.

If anything should ever be devised for similar work, and yet on a still smaller scale, it will have to border on the miraculous.

In less than three minutes everything could be disconnected and cleared right away from the room, and again be re-adjusted ready for work.

Such an important feature as this is not likely to be overlooked by the military authorities.

The station at the Needles' Hotel is similarly equipped, excepting that the pole is somewhat higher than that at Bournemouth.

From what has been stated in another chapter, it will be seen that the transmitting apparatus at Bournemouth must be in tune, or syntony, with the receiver at the Needles. Similarly, the transmitter at the Needles must be in tune with the receiver at Bournemouth.

The keenest interest will necessarily be taken in Signor Marconi's experiments between the Isle of Wight and Cherbourg, which are likely to be undertaken very shortly. The straight-line distance is about sixty miles.

Experiments that were looked upon twelve months ago as impossible are now accomplished successfully.

The compactness of the apparatus required is almost as surprising as the wonderful performance itself.

Appendix

A. INDUCTION AND CONDUCTION.

WHEN a current of electricity is set up in a wire, immediately magnetism is present around that wire; and if another wire be brought within the influence of the electrically-charged wire, it, too, will be electrically charged by induction if the two wires be placed parallel to each other.

In conduction, wires or other tangible means of communication—such as water, damp air, or the earth—must exist between the electrically-charged wire and the wire or other body to be influenced.

B. CURRENTS, VIBRATIONS, AND WAVES.

The long pipe of a speaking-tube will serve to illustrate the effect known as a ' current.'

The person speaking into the tube does not cause a current of air to pass from one end to the other. The act of speaking causes sound waves in the air, which impinge upon the ear, but there is no stream of air. Similarly, there are no currents of electricity forming a stream. Electrical waves or vibrations of the ether are produced.

In a stormy sea the dashing of the billows on the shore gives us the impression that a strong current, consisting of all the waves, is coming in with great velocity ; but a piece of wood floating on the waves some distance from the shore makes very little progress towards land, although it rises and falls with the waves.

C. INSULATED WIRES.

Just as light passes easily through glass, but not through iron, so electricity passes easily through some substances and not through others. Copper wires are covered with gutta-percha to prevent the escape of the electrical energy. The pipe of a speaking-tube will illustrate its functions. When a person speaks into a tube the sound-waves

are prevented from diverging in all directions
until they have reached the other end.

D. The Methods of James B. Lindsay.

The annual report of the British Association for 1859 contains the following brief
account of Lindsay's remarkable methods of
signalling without wires :

'The author has been engaged in experimenting on the subject, and in lecturing on
it in Dundee, Glasgow, and other places,
since 1831. He has succeeded in transmitting signals across the Tay and other sheets
of water by the aid of the water alone as a
means of joining the stations.

'His method is to immerse two large plates
connected by wires at each side of a sheet of
water, and as nearly opposite to each other
as possible. The wire on the one side from
which the message is to be sent is to include
the galvanic battery and the commutator or
other apparatus for giving the signal. The
wire connecting the two plates at the receiving station is to include an induction
coil or other apparatus for increasing the
intensity and the recording apparatus. The

distance between these plates he distinguished by the term "lateral distance. ..." He entered into calculations to show that two stations in Britain—one in Cornwall and the other in Scotland, and corresponding stations well chosen in America—would enable us to transmit messages across the Atlantic.'

E. THE RELAY.

Signor Marconi found that the current which could be started by the sensitive tube or contact was not sufficiently strong to work the tapper of the receiving instrument ; so, to overcome this difficulty, instead of obliging the current of the circuit which contains the sensitive tube to work the trembler and in-strument, he uses the said current for working a sensitive relay, which closes and opens the circuit of a stronger battery, preferably of the Leclanche type. This current, which is much stronger than the current which runs through the sensitive tube, works the trembler and the Morse inker, or the electric bell when Morse inker is removed.

He found that the relay should be one possessing small self-induction, and wound to a resistance of about 1,000 ohms. It should preferably be able to work regularly with a current of a milliampère or less.

F. The Filings in the Coherer.

The filings in the Marconi coherer are comprised of 96 per cent. of hard nickel and 4 per cent. of silver. The presence of the merest trace of quicksilver is necessary, but this is managed by amalgamating the faces of the silver plugs that form the terminals of the wires in the tube.

By increasing the proportion of silver the coherer becomes more sensitive, but there is a limit to the sensitiveness required. If too sensitive, it would obviously be affected by atmospheric electricity.

It also stands to reason that anything in excess of a mere trace of quicksilver would agglutinate the filings and prevent their instantaneous decoherence.

The particles of metals filed down should be of uniform size. The files used in filing the nickel and silver should be frequently washed and well dried. It will also be well to use them when warm.

The filings ought not to be fine, but rather as coarse as can be produced by a large and rough file.

G. The Capacities or Wings.

These are two copper or brass strips, about an inch in width, and about 12 inches long, the thickness being about the sixtieth of an inch.

It is considered essential that the coherer should be tuned to obtain the advantage supposed to result from electrical syntony. The rate of oscillation of electrical charges in the wings depends on their length, which has to be determined by experiment.

H. Choking Coils.

Of the uses of these Mr. Marconi says: 'Another improvement has for its object to prevent the high frequency oscillations set up across the plates of the receiver by the transmitting instrument, which should pass through the sensitive tube, from running round the local battery wires, and thereby weakening their effect on the sensitive tube or contact. This I effect by connecting the battery wires to the sensitive tube or contact, or to the plates attached to the tube through small coils possessing self-induction, which may be called choking coils, formed by winding in the ordinary manner a short length (about a yard) of thin and well-insulated wire round a core (preferably containing iron) 2 or 3 inches long.'

I. Points to be observed in the make of Coherers.

In sealing the ends of the glass tube a hydrogen and air-flame should be used so as to avoid oxidizing the silver blocks or the filings. The vacuum desirable is one of about one-thousandth of an atmosphere, obtained by a mercury pump.

If the sensitive tube has been well made, it should be sensitive to the inductive effect of an ordinary electric bell when the same is working from one to two yards from the tube.

To keep the coherer in good working order it is desirable, but not absolutely necessary, not to allow more than one milliampère to flow through it when active.

K. Another Method of De-Cohering.

A most ingenious and yet most simple adaptation of the combined functions of the coherer, relay and tapping apparatus is shown in Fig. 12.

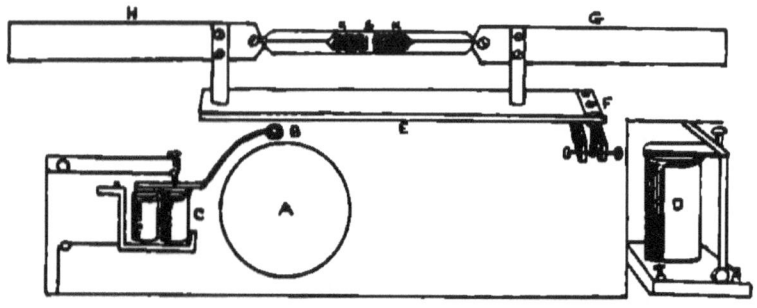

FIG. 12.—ANOTHER METHOD OF DE-COHERING.

It does not seem to be adapted to the appreciation of signals from greater distances than 60 or 80 feet.

It would, therefore, be useless for all practical purposes, except for lecture-hall demonstrations or entertainments.

This, it should be observed, is the writer's opinion from a superficial survey of the instrument only. At the same time one cannot avoid admiring some points in its construction.

First of all, the copper capacities are in line with the coherer.

There are no choking coils. This, however, is an indication that it is not expected to respond to long-distance signalling.

The hammer for direct and exclusive tapping is absent. This is obviated as follows : The two capacities or wings, and the intervening coherer are attached to a springboard E, the only support of which is at F. There is a relay at D. The contact breaker is at C. The bell is at A, and its hammer at B.

When energy reaches the wings G and H, it is conveyed to the filings in s, the current is completed, the hammer B strikes the bell, but its recoil shakes the springboard. The filings are thereby shaken and de-cohered, so that the current is broken. There can be nothing more simple, but whether it can be made equally sensitive with those made by Marconi is a matter which admits of doubt.

This apparatus is portable, compact, and attractive - looking, but apparently is only suited for demonstrations. The name of its maker does not transpire.

L. The Tapping or De-cohering Apparatus.

The trembler, or tapper, should be placed on the circuit of the relay, and should be similar in construction to that of a small electric bell, but having a shorter arm. Signor Marconi has used a trembler wound to 1,000 ohms resistance, having a core of good soft iron, hollow, and split lengthways, like most magnets used in telegraph instruments.

This tapper should be carefully adjusted. The blows should be directed from below, so as to prevent the filings from getting caked.

M. The Accident at Ferranti's.

About ten years ago an accident occurred at Ferranti's electric lighting depôt in Deptford, owing, it is thought, to a 'tremendous escape of electricity,' to put it popularly; one of the dynamos became connected to earth at night-time, with the result that all the South London telegraph plant became deranged and unworkable. The effect on electrical apparatus was felt in the central counties, and even on to Paris.

Referring to this, the result of an accident, Professor Thompson sees good foundation for the possibility of sending signals to great distances by conduction—earth or water being the conducting medium. Logically, and with excellent reasons, the Professor comes to the conclusion that if such things were possible without pre-arrangement, it is 'obvious that, by proper forethought and due expenditure of money on the requisite machinery, a telegraph without wires might be established between London and Paris, or, for that matter, between any two places.'

N. MILLER AND WOODS' RECEIVER.

This is a very compact instrument, essentially the same as one devised some years ago by Dr. Oliver Lodge. The coherer consists of a small ebonite cup containing nickel filings. Two wires supporting the cup pass into it and almost meet. The filings when cohered form the bridge from point to point of the wires. Two small rods, with a screw thread on them, and connected to the electric bell, play over the wires like a fiddle-bow over the strings. The complete apparatus does not occupy one cubic foot of space.

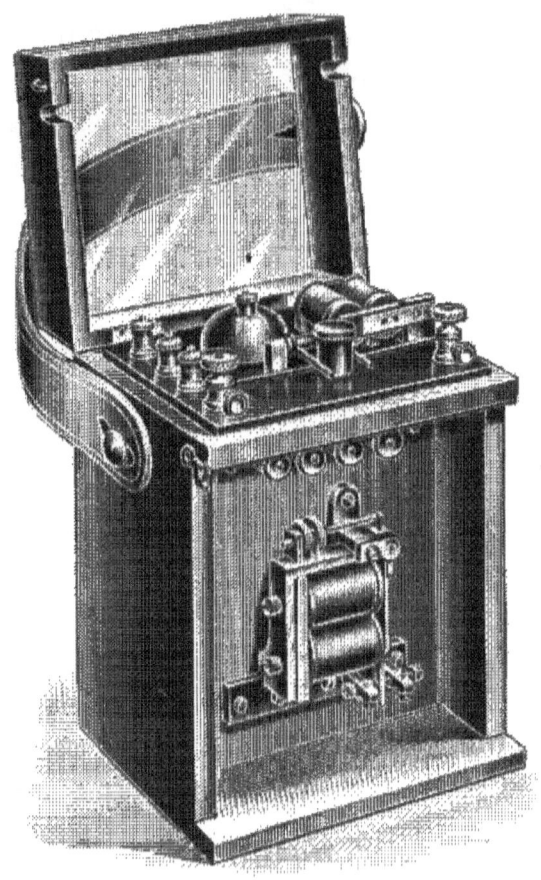

MILLER AND WOODS' RECEIVER FOR WIRELESS TELEGRAPHY.